# Snow White and the Seven Dwarfs

Illustrated by Eric Kincaid

Once there was a wicked Queen who
had a magic mirror. Every day she
would say,
"Mirror, mirror on the wall,
Who is the fairest of them all?"
Every day the mirror would reply,
"You, oh Queen,
Are the fairest in all the land."

One day, the Queen asked,
"Mirror, mirror on the wall,
Who is the fairest of them all?"
And the mirror replied,
"You, oh Queen, are very fair,
But Snow White is the fairest in
all the land."
Instead of her
own face looking
from the mirror
the Queen saw the
face of her step-
daughter. She was
VERY angry.

She sent for her huntsman. "Take Snow White into the forest and kill her," she said.

The huntsman loved Snow White. "I cannot kill you," he said. "But I cannot take you home either. You must stay here, in the forest."

The huntsman returned to the palace alone. He told the Queen he had killed Snow White.

Snow White wandered through the dark forest. She did not know where to go, or what to do. Presently she came to a little house. "Perhaps the people who live here will help me," she said. She knocked at the door. There was no reply, so she peeped inside.

What an untidy house it was. Every
thing in it was smaller than usual.
And there seemed to be seven of
EVERYTHING. Seven chairs. Seven
beds. Seven spoons. Seven plates.
Seven mugs. Seven of EVERYTHING
. . . except tables. There was just
one table.

"Perhaps children who have lost their mother live here," said Snow White. "I will tidy the house for them."

She swept, and dusted and cleaned and polished. She had plenty of helpers.

There was a diamond mine on the far side of the forest. It was worked by seven dwarfs. The very same dwarfs who lived in the house Snow White had found. They were on their way home. What would happen to Snow White now?

"Who . . . who are you?" asked Snow
White. "This is our house," said the
seven dwarfs all together. "We work
in the diamond mine on the far side
of the forest. Do not be afraid, we
will not hurt you. Just tell us
what are YOU doing in OUR house?"

Snow White told the dwarfs what had happened.

"You can stay here with us. We will look after you." said the dwarfs.

"And I will look after you," said Snow White. "I will cook and sew and clean for you."

And she did. And they were all very happy.

But the dwarfs
were afraid the
wicked Queen would
come looking for
Snow White one day.
"Do not answer
the door to
anyone," they said,
whenever they went
to the mine.

They were right to be worried.

"Mirror, mirror on the wall,
Who is the fairest of them all?"
asked the Queen.
"You, oh Queen, are very fair,
But Snow White, who lives in the
forest with the little men, is
the fairest in all the land."
The Queen was VERY ANGRY
INDEED. "I will kill Snow White
myself!" she said.

The Queen disguised herself as a pedlar. She filled a basket with apples then went into the forest. She waited until the dwarfs had gone to the mine, then she knocked at the door of the little house. "I have nothing to fear from a pedlar," said Snow White. And though the dwarfs had warned her not to open the door, she did.

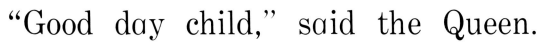

"Good day child," said the Queen. "Would you like one of my apples?" "Yes please . . ." said Snow White.

The Queen gave Snow White the reddest apple in the basket. It was a special apple. A VERY special apple. The Queen had put a spell on it.

"Take a bite . . ." said the Queen. Snow White took just one bite from the apple and fell to the floor. "Ha! Ha!" laughed the Queen, and she threw off her disguise. "Snow White is dead. Now I am the fairest in all the land!"

When the dwarfs came home they found Snow White lying on the floor. They found the apple, with one bite taken from it, lying beside her. "The wicked Queen has been here," they said sadly. "Snow White is dead!"

The dwarfs built a special bed for Snow White in the forest. The birds and the animals kept watch around her.

One day a prince came riding by and saw her lying there.
"Please let me take her home," he said.

As the Prince lifted Snow White
onto his horse she opened her eyes.
The piece of magic apple had fallen
from her throat. The spell cast by
the wicked Queen was broken.
"Snow White is alive!" shouted the
dwarfs. "Hoorah! Hoorah!"

Once more, the Queen asked the magic mirror who was the fairest of them all.
The mirror replied,
"You, oh Queen, are very fair,
But Snow White, the Prince's bride, is the fairest in all the land."
The Queen was so angry, she flew into a rage and died herself. And so Snow White was safe at last.

All these appear in the pages of
the story. Can you find them?

Queen

mirror

Snow White

huntsman